Part 1
Cars and Trucks

THINGS THAT GO

Part 3
Diggers and Cranes

Part 2
Planes and Helicopters

Part 4
Ships and Boats

Cars and Trucks Contents

Cars 2
At the controls 4
Fast, faster, fastest 6
Hot rods and rally cars 8
Tough cars for rough places 10
To the rescue 12

Trucks 14
Artics 16
Loading and unloading 18
Tankers 20
Trucks for special jobs 22
Did you know? 24

zigzag

Fun Fact

A car has over 14,000 pieces.

Types of car

Convertibles have tops that fold back.

Estate cars have a large space in the back for luggage.

Saloon cars have two or four doors and a boot.

Hatchbacks have a door at the back that opens upwards.

At the controls

Cars are complicated machines but the controls to drive them are quite simple.

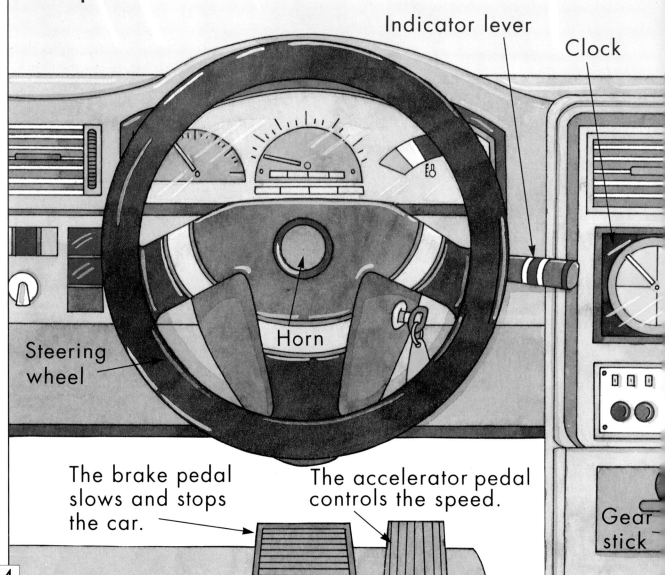

Indicator lever

Clock

Steering wheel

Horn

Gear stick

The brake pedal slows and stops the car.

The accelerator pedal controls the speed.

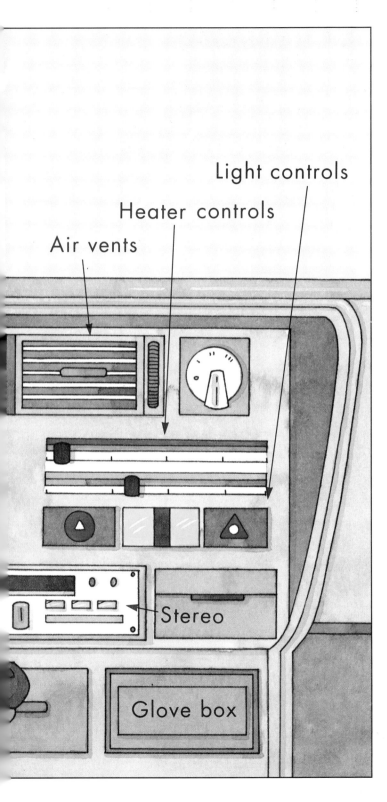

Control panel

The dials and lights on the control panel give the driver important information.

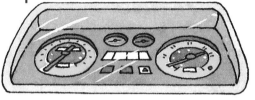

The **low level fuel warning light** comes on when the car needs more petrol.

The **temperature dial** warns when the engine is overheating.

The **speedometer** shows how fast the car is going.

Fast, faster, fastest

Sports cars have powerful engines and streamlined shapes so that air can pass over them easily.

Turbo engine

Streamlined shape

Spoiler

Fun Fact

Speeding racing cars go so fast that sometimes they almost take off! Aerofoils (spoilers) attached to the car act like upside-down wings, keeping the tyres on the ground.

Formula One racing cars are built for one thing – to win races! There is just enough room inside for the driver.

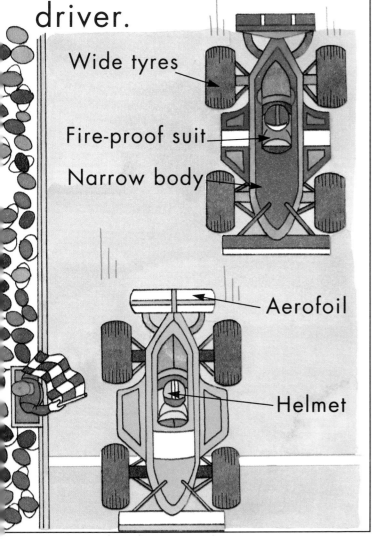

Built for speed

Slicks Smooth wide tyres called slicks give racing cars extra speed.

Light bodies Racing cars have bodies made of extra-light material.

Powerful engines and lots of gears drive racing cars at speeds up to 400 km/h (250 mph).

Hot rods and rally cars

Hot rods and dragsters are sprinters. They run in drag races. The winner is the fastest to the finishing line – only 1/4 of a mile away from the start!

Hot rods

Spoiler

Cage to protect the driver.

Dragster

Drag strip

Slick-tread tyres

Rally cars are long distance racers. They run in races that last for days, or even weeks or months.

Dragsters go so fast that they need **parachutes** to help slow them down.

Rally drivers have a **navigator** in the car to tell them which way to go.

Fun Fact
The longest rally race ever run was from London to Sydney – 31,107 km (19,329 miles).

Tough cars for rough places

Tough cars, called off-road vehicles, are built to bump over rocky ground and splash through water. They can go up steep hills and keep a tight grip on downhill slopes.

On the farm

Outdoor pursuits

On safari

Fun Fact
Some car headlights have small washers and wipers to keep them clean.

Special features
Ground clearance

The bottom of the car is high to clear rocks and bumps.

Tyre tread

Deep tyre patterns, called tread, grip slippery or sandy ground.

Four-wheel drive

All four wheels are powered by the engine. If any of the wheels gets stuck, the other wheels can still work.

To the rescue

Police cars, fire engines and ambulances are fitted with special equipment to deal with emergencies. Sometimes they all work together.

Fun Fact

FIRE, POLICE and AMBULANCE are sometimes written like this so that drivers can read them through their rear view mirrors.

Doing their jobs

Emergency vehicles use blue **flashing lights** and loud **sirens** to warn other drivers to pull over so they can get past.

An **aerial ladder platform** can rescue people from the top of buildings, down cliffs and under bridges.

Ambulances have **medical equipment** inside to give patients treatment on the way to hospital.

Trucks

Trucks carry and deliver loads all over the country and abroad. They have different shaped bodies for the kinds of loads they carry.

Door mirrors set wide for extra vision.

Box body
A removal truck carries furniture in its box body.

Curtain-sider
The curtains keep the load dry and in place.

Drop-sider
The sides of this truck can be dropped down to unload the scaffolding.

Clips

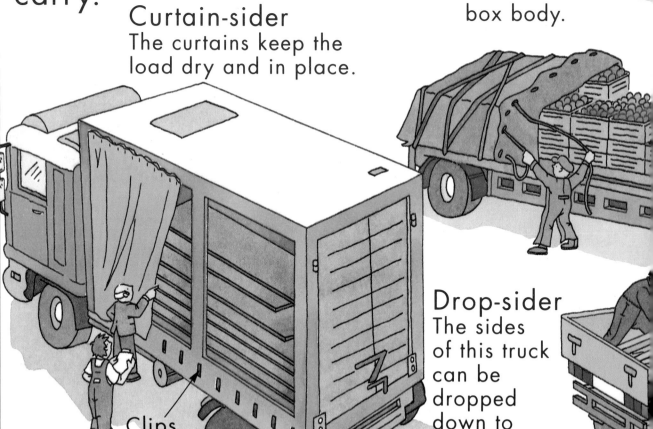

Roll-up door

Flatbed body
It is easy to reach the load once the sheeting and ropes are untied.

Truck parts

Straight trucks have two sets of wheels. Sometimes the back wheels are in pairs.

Chassis The chassis is the frame of the truck.

Axle Each set of wheels is on an axle.

Artics

Articulated trucks, or artics, have two parts – the tractor at the front and the trailer at the back. Tractors can drive without a trailer, but trailers cannot go anywhere without a tractor.

Air deflector

Horns

Airlines

Trailer

Engine

Lights

Fun Fact

Heavy trucks often have more than six pairs of wheels.

Cab

Tractor

Cab tilt The cab can be tilted for the driver to check the engine.

Night cab Long distance drivers sleep in a night cab.

Air lines are plastic pipes that carry air for the brakes in the trailer.

Loading and unloading

Different loads need special trucks to move them.

A loading machine fills the tipper with buckets of earth.

Some trucks are fitted with cranes for lifting their heavy loads.

A fork-lift truck can load boxes right into the top of another truck.

The middle and top platforms of a car transporter tip to make ramps for driving cars on and off.

On and off

Crane The crane can lift, swing round and lower its head.

Forks Fork-lift trucks have a two-pronged fork which slots into special wooden loading pallets.

Tipper A rod behind the driver's cab opens like a telescope to tip the truck up.

Tankers

Tankers carry liquid loads such as petrol, chemicals and milk. They can also carry powder and granules.

Petrol tanker
Petrol is pumped from the petrol tanker into tanks under the forecourt of the service station.

Milk float

Weighbridge

Gauge

Delivery pipe

Underground storage tank

Milk tankers
Milk tankers collect milk from farms and deliver it to the dairy.

Petrol pumps

Forecourt

Loads

Dry loads Tankers can tip up to empty dry loads.

Liquid loads are delivered through a pipe.

Dangerous loads have warning stickers. In an accident, a code tells the rescue workers what is in the tanker.

Trucks for special jobs

Some trucks are fitted with machinery for doing special jobs.

Road sweeper
A brush whirls round cleaning and sweeping.

A giant vacuum pipe sucks up the dirt and leaves.

Snow plough

Strong headlights are needed to work in blizzards.

A spreader swirls salt and grit on to the road to stop it icing up again.

A blade scrapes and pushes snow and ice off the road.

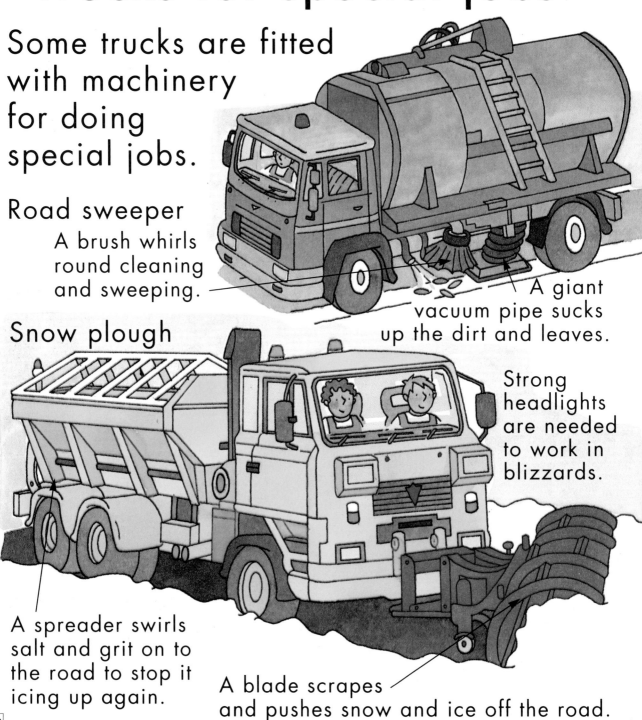

Refuse truck

The back section lifts up and the rubbish is tipped out at a disposal site.

Rotating blades crunch up the rubbish.

Plastic bins are emptied into the truck.

Ultra-heavy hauler

These trucks carry enormous loads. An extra truck is often needed to push from behind or to pull from the front.

A **police escort** with flashing lights and sirens travel with the load to warn other drivers.

Did You Know?

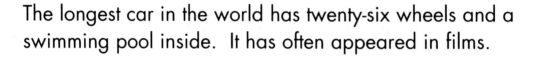

The longest car in the world has twenty-six wheels and a swimming pool inside. It has often appeared in films.

Some cars can travel on sea as well as over land. They are called amphibious cars. In 1958 a man called Ben Carlin completed a journey around the world in an amphibious car.

Some ambulances called jumbulances can carry up to 44 people. They are used for big groups of sick people.

Part 2
Planes
and Helicopters

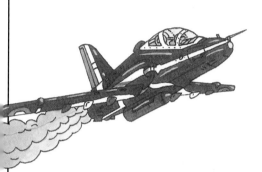
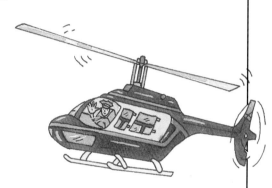

Contents

Flying 26
In the cockpit 28
Airliners 30
Concorde 32
Special planes 34
Aerobatics 36

Small planes 38
Gliders 40
Helicopters at work 42
Search and rescue 44
Skycranes 46
Did you know? 48

Flying

Planes can fly because they have engines and wings. Engines push the plane along the runway. Air rushes over the wings and lifts the plane into the air.

Fun Fact

In 1903 Orville Wright made the first controlled flight in history. It lasted for only 12 seconds!

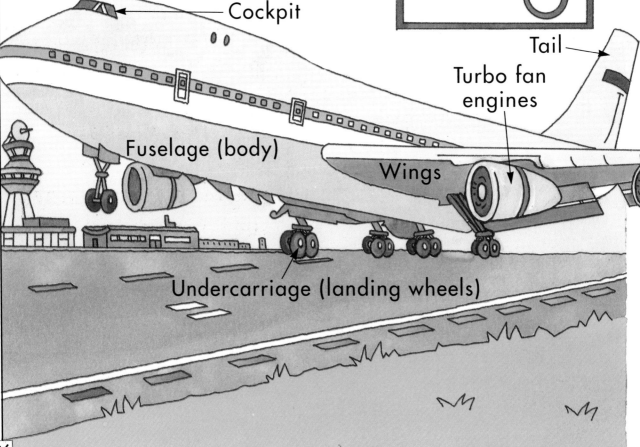

- Cockpit
- Tail
- Turbo fan engines
- Fuselage (body)
- Wings
- Undercarriage (landing wheels)

Helicopters do not need runways. Their engines drive rotors that spin and lift the helicopter straight up into the air.

Main rotor
Tail
Tail rotor
Cockpit
Skids

Engines

Propellers Some planes have propellers. The engines make the propellers spin round on the front of the plane.

Jets The fastest planes have jet engines. Jet engines suck in air and push it out behind them.

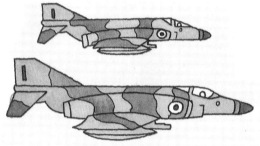

In the cockpit

The pilot controls the plane from the cockpit. In airliners this is usually called the flight deck. Lots of screens and dials give the pilots all the information they need to fly the plane.

Microphone

Runway

Pilot (Captain)

Fun Fact
Airline pilots and co-pilots are given different meals. If the food makes one of them ill, the other can take over!

Earphones

Co-pilot

Special instruments

Control stick
The pilot moves the control stick backwards, forwards and sideways to control the plane.

Autopilot This is a computer that controls a plane in flight. Pilots control the take-off, but the autopilot can do most of the work during the flight, and even land the plane!

Airliners

Each year, millions of passengers are carried all over the world in airliners. The largest are jumbo jets that carry more than 500 passengers. Small airliners carry 20 to 30 people.

Boeing 747 (jumbo jet) — Tail plane, Tail, Hold, Engines

Fun Fact
The tail of a jumbo is as high as a six storey building.

Airliner shapes

Four engines Some airliners have four engines, two on each wing.

Three engines Some have three engines, one on each wing and one on the tail.

Two engines Some have two engines, on the wings or on either side of the tail.

Concorde

Concorde is the world's fastest airliner. It is supersonic, which means it flies faster than the speed of sound.

Streamlined shape

Passenger cabin

Delta-shaped win

Concorde facts

Concorde has four engines and can fly at more than 2,000 km/h.

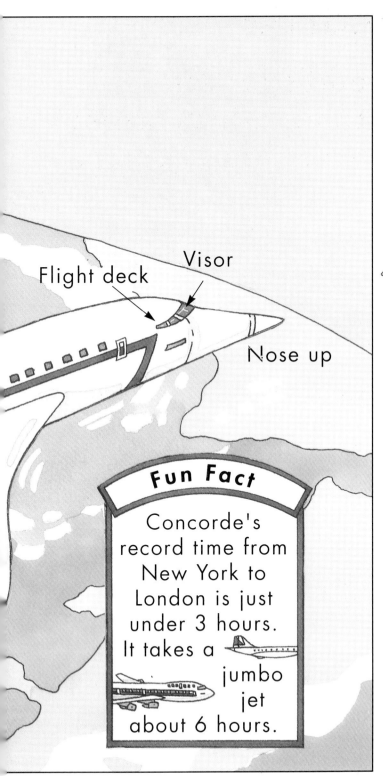

Its visor and nose can be dropped down to let the pilot see the runway clearly.

When Concorde reaches the speed of sound, you can hear a double bang.

Fun Fact

Concorde's record time from New York to London is just under 3 hours. It takes a jumbo jet about 6 hours.

Special planes

Here are some unusual planes from around the world.

The Harrier jump jet can take off straight up into the air like a helicopter. Then it swivels its jet engines round and flies forward.

Floats

Seaplanes have floats instead of wheels. They can land on water.

Jet engines

The Galaxy military plane is huge. It can carry heavy loads of tanks, trucks and soldiers.

Refuelling

Small fighters cannot carry much fuel.

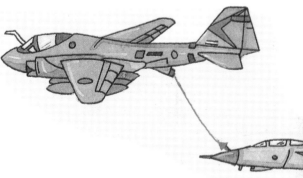

On long flights they refuel from special refuelling planes. The big plane has long pipes coming from its fuel tanks. The small plane fits a probe into a pipe and takes in fuel.

Aerobatics

Pilots control a plane's movement by changing the position of flaps on the tail and wings. Display teams, like these Red Arrows, can roll, loop and spin their planes. This is called aerobatics.

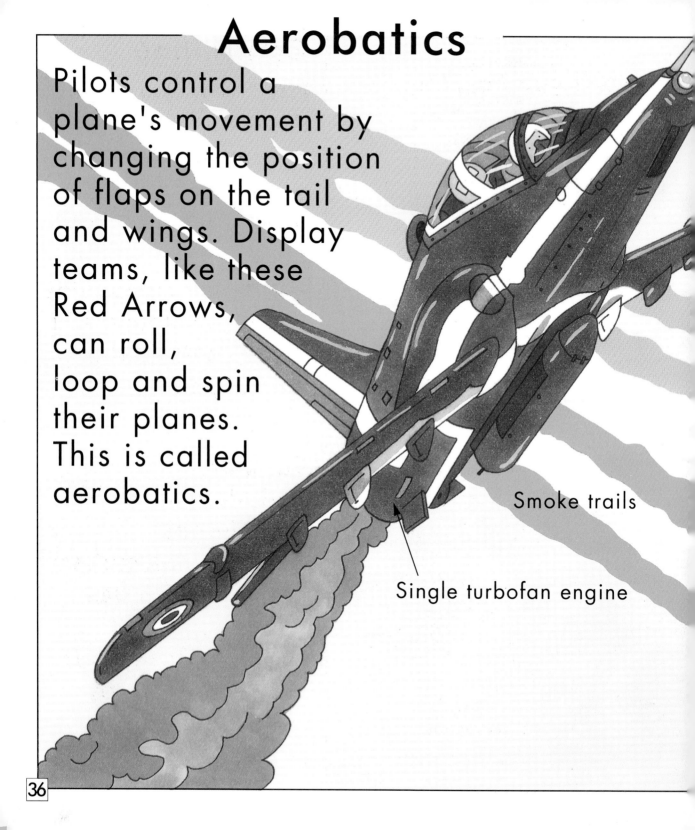

Smoke trails

Single turbofan engine

Fun Fact

People used to perform in flying circuses by standing on the wings of planes while they flew!

Formation

Controlling

Pitch is the up and down movement. It is controlled by the elevator flaps.

Yaw is sideways movement. It is controlled by the rudder.

Bank or Roll Moving the ailerons on the wings makes a plane roll over.

Turning is done by banking.

Small planes

Small planes can be fun for sport or they can do special kinds of work.

The tiny Pitts Special biplane was built to do extraordinary aerobatics.

Pitts Special

Microlight

Microlights are hang-gliders with engines. They are used for sport.

The bubble-shaped cockpit of the Optica gives pilots a clear view of the ground. It is used by some police forces.

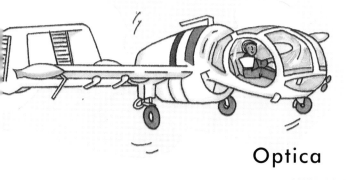

Optica

Around the world

Small planes are used around the world for special jobs.

Landing on ice

Flying doctors

Piper Aztec

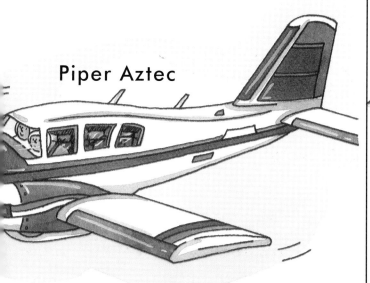

Small propeller planes like this Piper Aztec can carry two to six people.

Crop spraying

Gliders

A glider has no engines to power it. An aeroplane, truck or a winch has to tow it along until it is going fast enough to fly on its own.

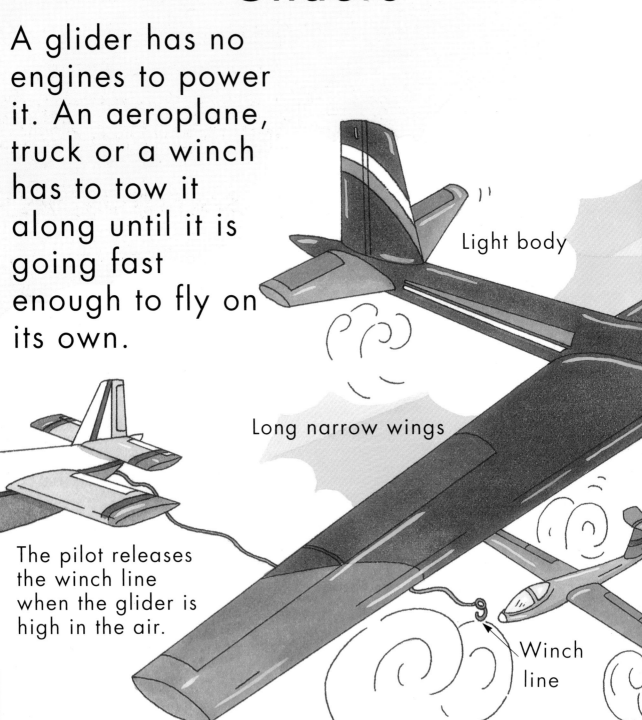

Light body

Long narrow wings

The pilot releases the winch line when the glider is high in the air.

Winch line

Hang gliders

The pilot can be launched by being towed by a winch, or by running down a hillside.

The pilot is strapped in beneath the fabric wings. He or she uses a control bar and body movements to steer the glider.

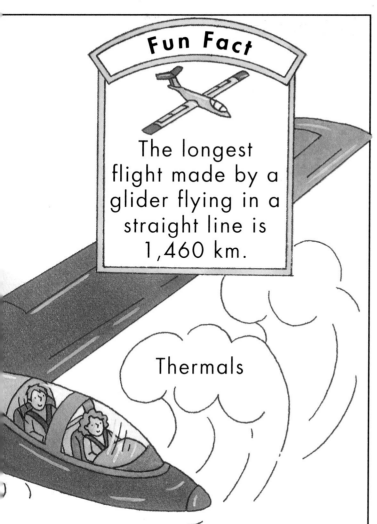

Fun Fact

The longest flight made by a glider flying in a straight line is 1,460 km.

Thermals

Streamlined shape

Rising currents of warm air (called thermals) keep the glider in the air.

Helicopters at work

Helicopters make good working machines because they can land and take off in very small spaces. They can also move up, down, backwards, forwards and sideways, and even hover in one place.

Passenger helicopters fly between airports, or take crew and supplies to oil platforms.

Movements

Up and down When the pilot tilts the rotor blade upwards, the helicopter lifts into the air.

Forwards and backwards Tipping the rotor forwards makes the helicopter fly forwards. When the rotor is tipped back, it flies backwards.

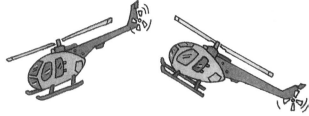

Tipping the rotor sideways makes it fly to the left or the right.

Fun Fact

The first helicopter flew in 1907. It lifted 1 metre off the ground for about 20 seconds!

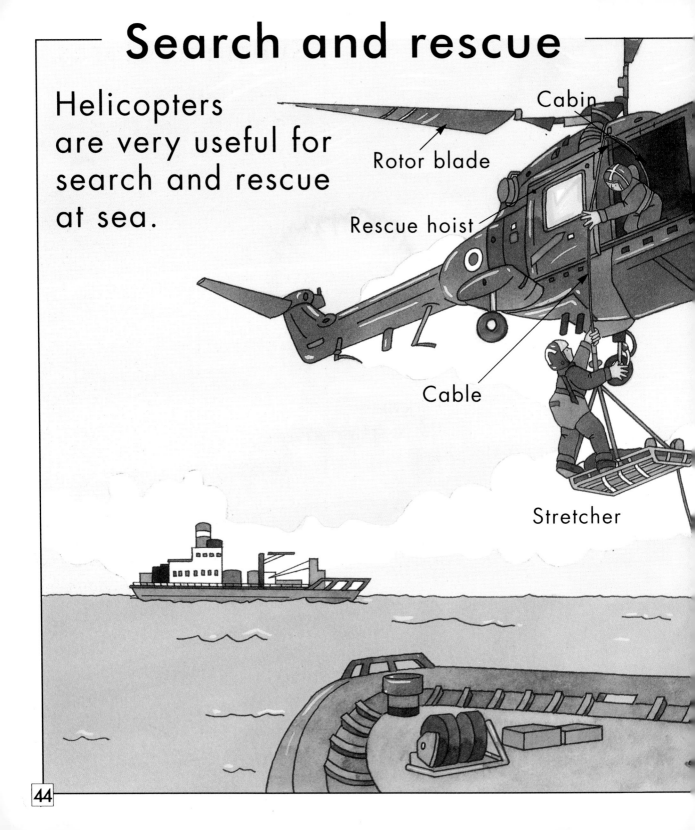

Search and rescue

Helicopters are very useful for search and rescue at sea.

Cabin

Rotor blade

Rescue hoist

Cable

Stretcher

One crew member goes down on a cable and puts a special collar round the person being rescued. Then the cable is winched up into the helicopter with both people attached.

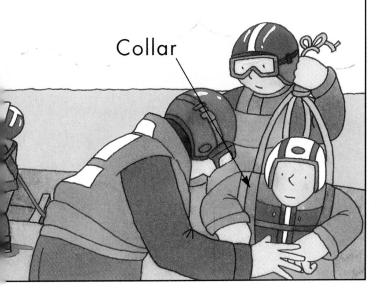

Collar

Special features

Helicopters with spotlights are used by the police.

Skycranes

Big, powerful helicopters are used as skycranes. They can lift and carry heavy loads to places that are very difficult to reach.

Sliding doors

Cargo hook

Main wheels

Fun Fact

The largest helicopter ever built was the Mil Mi-12. It could lift the weight of nearly 6 elephants!

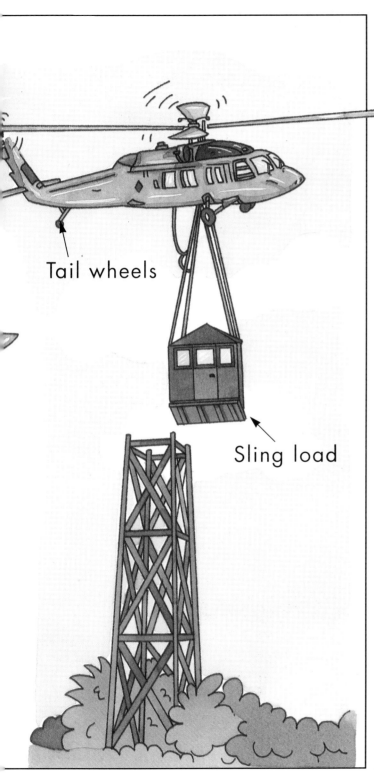

Tail wheels

Sling load

Special features

The Chinook helicopter has twin rotors that turn in opposite directions.

Safety The big rotors turn safely, high above people's heads.

Some helicopters have an opening in the cabin floor so the crew can watch the load below.

Did You Know?

A flying boat is a special plane which can travel on water as well as land. During the Second World War they were used for hunting down submarines and rescuing people.

Some planes use the Sun's energy to fly. They are called solar powered planes. "The Solar Challenger" was the first of these solar operated planes to fly across the English Channel in 1981.

Jumbo jets can carry more than 500 people at a time. However, during very bad winds in Australia one plane carried out more than 670 people.

Part 3

Diggers
and Cranes

Contents

Backhoe loader 50
Excavators 52
Giant excavators 54
On the farm 56
Road building 58
Mining 60

Tunnelling 62
Tower cranes 64
Travelling cranes 66
Cranes in factories 68
Floating cranes 70
Did you know? 72

Backhoe loader

The backhoe loader is two machines in one. It does some jobs with the hoe at the back and other jobs with the loader at the front.

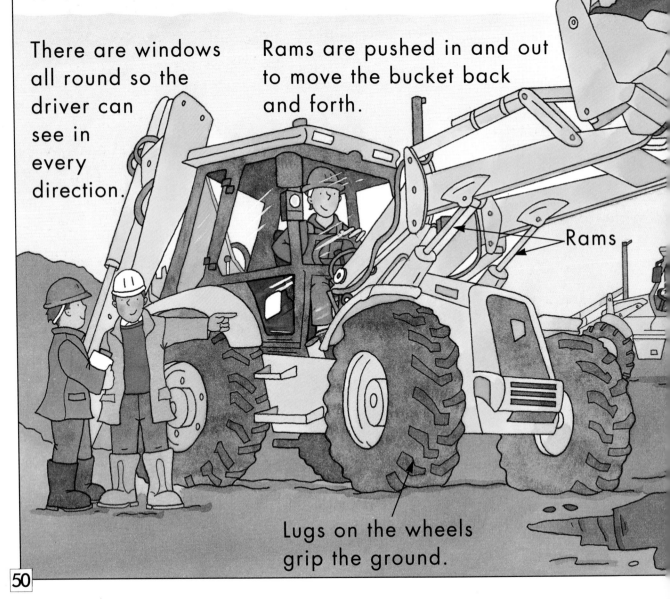

There are windows all round so the driver can see in every direction.

Rams are pushed in and out to move the bucket back and forth.

Rams

Lugs on the wheels grip the ground.

Bucket

Fun Fact

A big backhoe loader could pick up a mini excavator in its bucket!

The driver's seat can turn round to face the front or the back.

Arm

Backhoe

Stabilizers keep the machine steady while it is digging.

Special features

The loader bucket has strong metal teeth for digging.

The bucket can also split in half to grab piles of soil.

The backhoe arm can lift up and down, reach out and pull in.

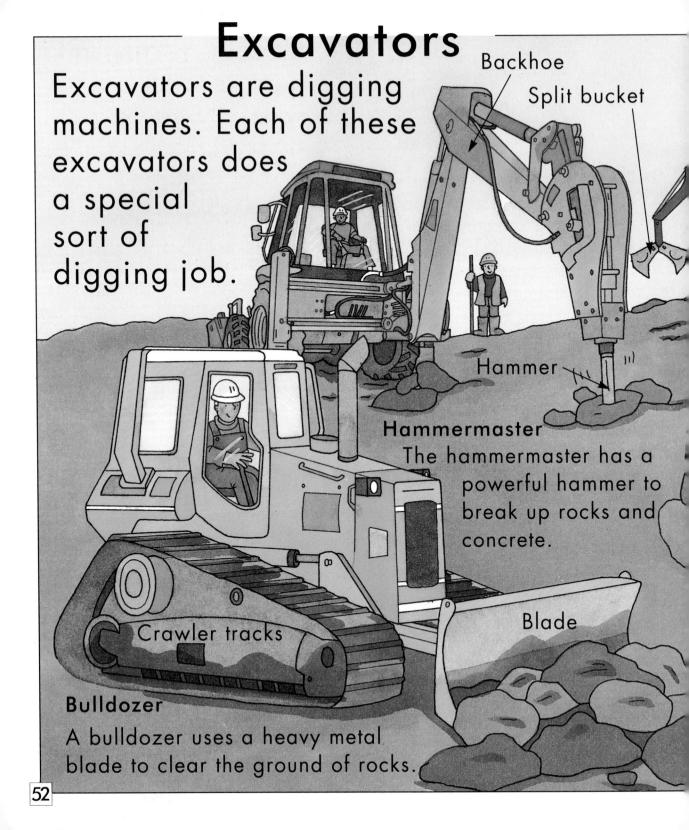

Excavators

Excavators are digging machines. Each of these excavators does a special sort of digging job.

Backhoe

Split bucket

Hammer

Hammermaster
The hammermaster has a powerful hammer to break up rocks and concrete.

Crawler tracks

Blade

Bulldozer
A bulldozer uses a heavy metal blade to clear the ground of rocks.

Split bucket

A split bucket opens and shuts like a mouth to bite into the earth.

Shovel

This special backhoe can swivel round to dig at the back, front or on either side.

Mini excavators and mini loaders are specially made to work in small spaces.

Mini skiploader

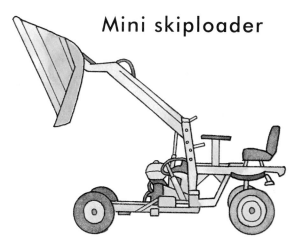

Mini excavator
with bucket and crawler tracks

Giant excavators

Rocks and stones used for building are dug out of quarries. The biggest excavators of all do this work.
They are called dragline excavators.

Fun Fact
A dragline bucket is so big that a car could drive into it!

The enormous bucket scoops up huge amounts of rock.

Draglines are the wires that drag the bucket along the ground so that it fills up with rocks.

Some excavators can walk backwards and forwards on big metal pieces called shoes.

Quarrying cuts huge holes in the earth. Dragline excavators can also be used to fill the holes in again!

On the farm

Tractors with powerful engines and big wheels are very useful on farms. They can pull many different kinds of farm machines. All these machines get the ground ready for sowing seeds.

Tractor

Plough
A plough has sharp blades called ploughshares. They dig into the soil and turn it over.

Ploughshares

Roller
A roller smooths the soil and makes the field level.

Harrow
A harrow has rows of metal discs that cut the soil up into small pieces.

← Metal discs

Seed drill

This is used to plant seeds in fields.

Hopper
Spikes
Furrows →

First the drill makes lines, called furrows, in the soil.

Seeds pass out of the hopper, down the drills into the furrows.

Spikes pull soil over the seeds.

Road building

The ground has to be cleared and flattened before a new road can be built. All these machines work together to do the job.

Compactor
Compactors drive up and down flattening the ground.

Blade

Metal feet

Blade

Grader
Graders use a blade to smooth over the surface.

Wheels

Scraper
Scrapers scrape away small bumps.

Blades

Bulldozer
Bulldozers clear big obstacles out of the way.

The scraper blade is lowered to the ground. A trap door lifts as it collects a load. When full the load is pushed out from the back.

Compactors have heavy metal wheels covered in spikes called feet. They push down into the soil.

Graders have long blades. They spread out small stones in an even layer.

Mining

Mining machines dig tunnels and cut coal, or metals such as gold, from deep under the ground.

Roadheader
A roadheader has spiky blades so it can bore tunnels through rock and earth.

Rock falls onto these wheels. The wheels turn and carry the rock away from where the machine is cutting.

Coal face cutter
This coal face cutter moves along the tunnel, cutting off coal from the wall of the mine.

Fun Fact
Once all the coal is removed, the tunnel is allowed to collapse.

Roof supports strengthen the roof and walls of the tunnel so they do not fall in.

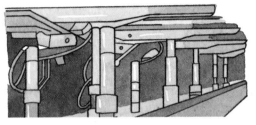

A **conveyor belt** carries the coal out of the tunnel.

Water sprays out of the coal cutter to damp down the coal dust.

Tunnelling

This giant tunnel-boring machine is called a TBM. TBMs were used to cut through rock to build the Channel Tunnel under the sea between England and France.

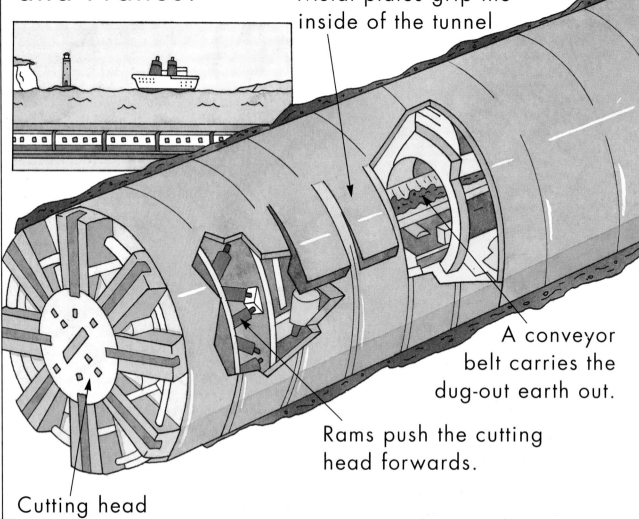

Metal plates grip the inside of the tunnel

A conveyor belt carries the dug-out earth out.

Rams push the cutting head forwards.

Cutting head

Concrete segments cover the inside of the tunnel.

The TBM control cabin has television cameras and computers.

Cutting head

This works like a drill. **Rams** push it against the rock. A big electric **motor** turns it round.

The huge cutting head can cut 12 cm of new tunnel in one minute.

Fun Fact

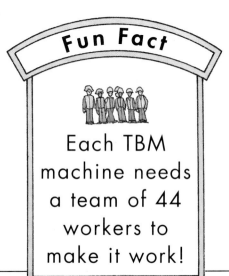

Each TBM machine needs a team of 44 workers to make it work!

Tower cranes

Cranes lift heavy things. Tower cranes work on building sites. They can lift heavy loads up to the top of the tallest buildings.

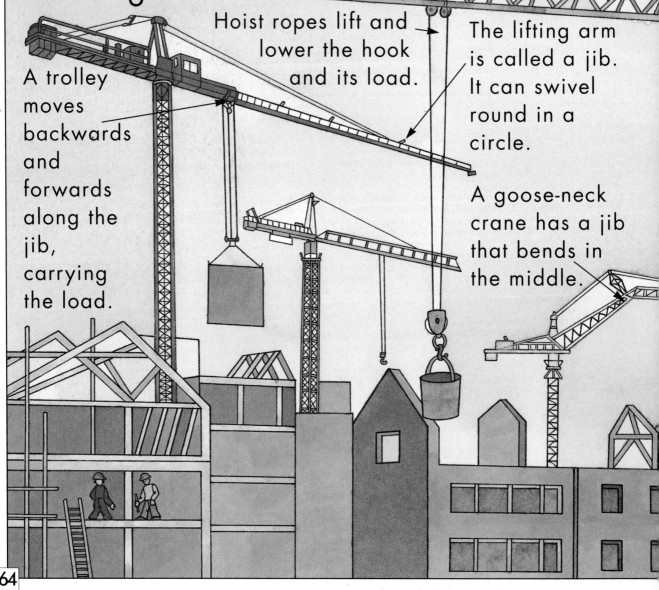

A trolley moves backwards and forwards along the jib, carrying the load.

Hoist ropes lift and lower the hook and its load.

The lifting arm is called a jib. It can swivel round in a circle.

A goose-neck crane has a jib that bends in the middle.

There is a big block of concrete at one end to stop the crane tipping over when it is picking up a heavy load.

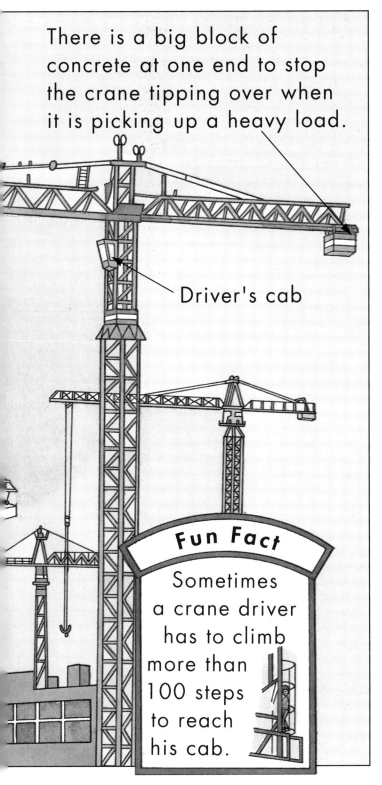

Driver's cab

Fun Fact

Sometimes a crane driver has to climb more than 100 steps to reach his cab.

Hoists

Hoists are used for lifting.

A **beam hoist** can lift a window frame into place.

This hoist carries people and materials in a cage to the top of a tall building.

Access platforms lift workers into the air in a safety cage.

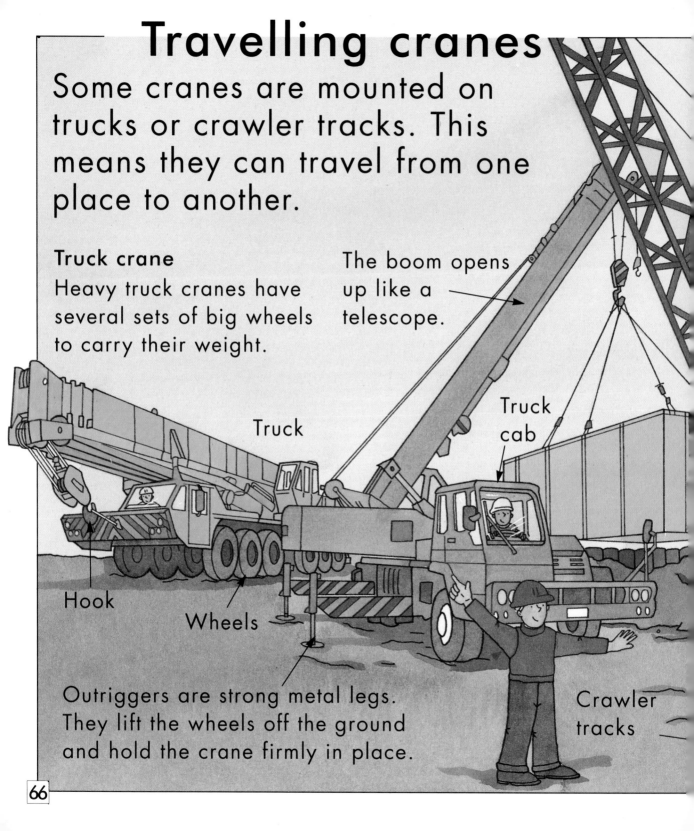

Travelling cranes

Some cranes are mounted on trucks or crawler tracks. This means they can travel from one place to another.

Truck crane
Heavy truck cranes have several sets of big wheels to carry their weight.

The boom opens up like a telescope.

Truck

Truck cab

Hook

Wheels

Outriggers are strong metal legs. They lift the wheels off the ground and hold the crane firmly in place.

Crawler tracks

Fun Fact

Truck cranes can lift loads as heavy as 30 elephants!

Crawler crane

Crawler cranes can lift heavy loads and move along at the same time.

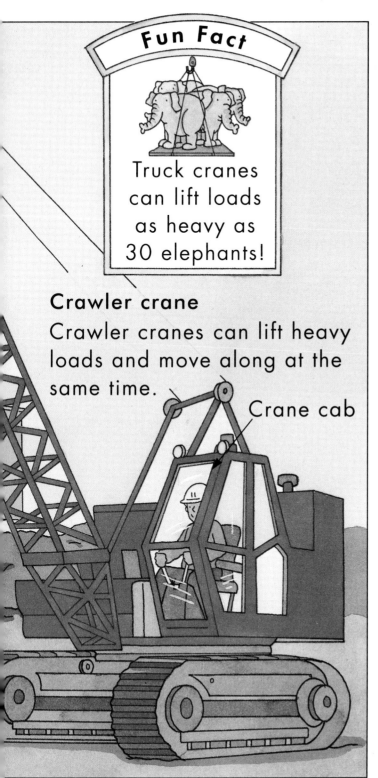

Crane cab

Crane controls

Hand levers and foot pedals are used to do these things:

Lift and lower the boom and the hook.

Swing the crane round.

The crane operator can see all round through the windows.

Cranes in factories

Travelling cranes are used in some factories. They are fixed above the factory floor and move backwards, forwards and sideways along rails.

This crane can lift a whole railway carriage.

The manipulator on the end of this crane lifts machinery off a conveyor onto a trolley.

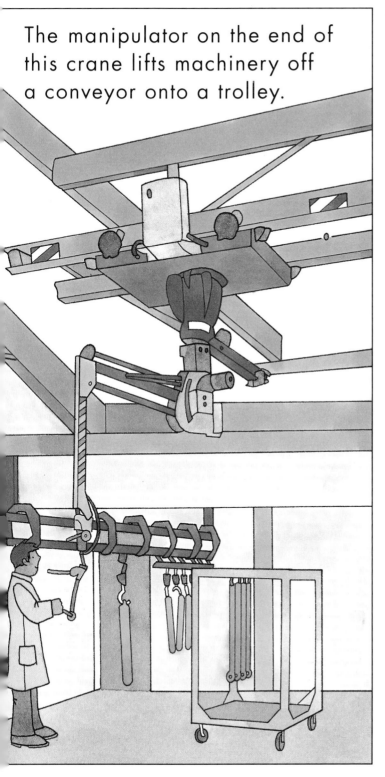

Special lifting jobs

This crane has been specially made to lift and carry a bucket of hot molten metal.

This crane lifts and carries metal by picking it up with a very strong magnet.

A frame with special lifting pads can pick up and carry a huge sheet of heavy glass.

Floating cranes

Floating crane vessels work at sea. This one is building an offshore oil rig. It has two booms that lift parts into place.

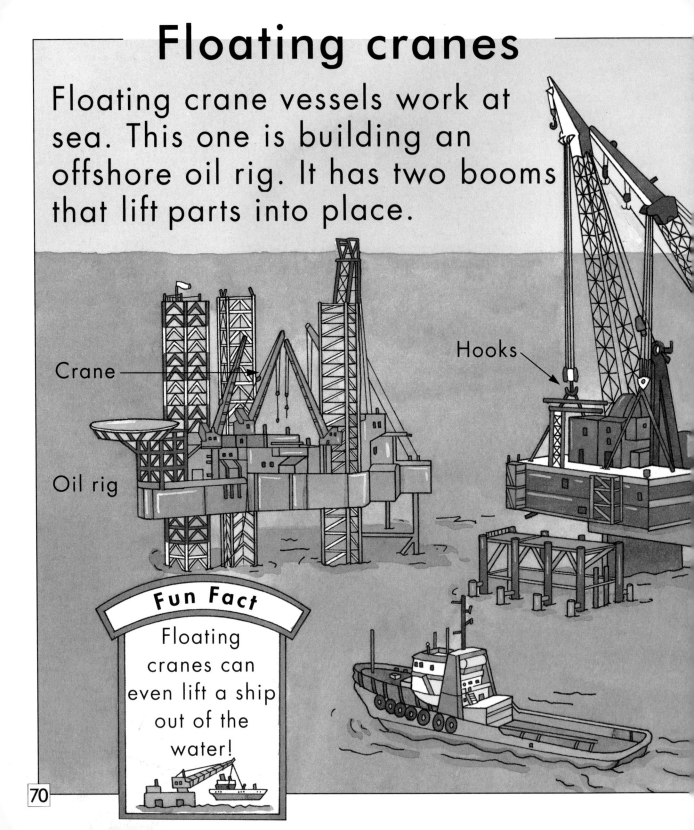

Crane

Hooks

Oil rig

Fun Fact

Floating cranes can even lift a ship out of the water!

About 350 people live and work on a floating crane vessel.

Boom

Helicopter

Helipad

Deck

There is a restaurant, cinema and hospital on board.

Huge **propellers** power the floating crane. It sails from one job to the next.

Helicopters carry workers and supplies to the oil rig and crane.

Smaller **floating cranes** are used to load and unload cargo ships when they are in harbour.

Did You Know?

Some of the biggest tractors in the world have wheels that are taller than grown men. One such tractor is called a Big Bud and is used for work on the long wheat fields of North America.

The biggest cranes are at sea and are used for lifting oil rigs.

The most powerful cranes in the world can lift loads as heavy as 500 double-decker buses.

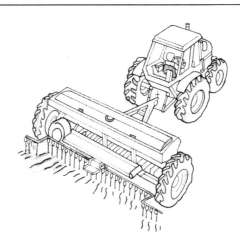

Part 4

Ships
and Boats

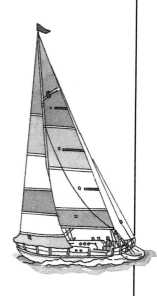

Contents

Afloat 74
Ferries 76
Fishing 78
Floating hotels 80
Cargo carriers 82
Warships 84
Ships for special jobs 86
Sailing 88
Speeding 90
Oars and paddles 92
Finding the way 94
Did you know? 96

Afloat

All ships and boats, from giant supertankers to tiny rowing boats, have the same names for their different parts.

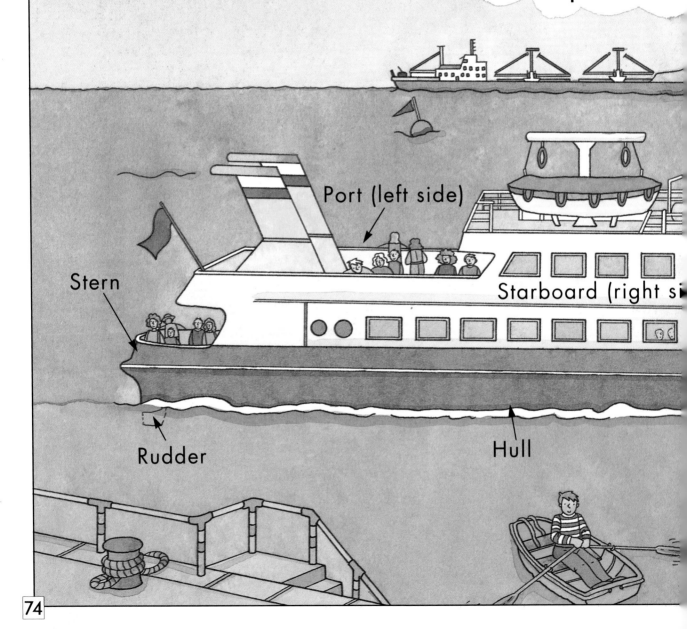

Fun Fact

Rich people who went to the Far East on the cool side of the ship were called **POSH** - because they travelled on the **P**ort side **O**ut and the **S**tarboard side **H**ome.

Bow
Anchor

Engines are used to drive ships and boats of all sizes.

Sails Sailing boats move along when the wind fills their sails.

Oars and paddles Rowers pull oars and paddles through the water to move their boats along.

Ferries

Ferries carry passengers, cars, trucks and sometimes even trains on short journeys from port to port.

Hovercraft
Hovercraft ride just above the waves on a cushion of air.

Ro-ro
Huge roll-on, roll-off ferries load and unload cars and trucks very quickly. They drive on and off through bow and stern loading ramps.

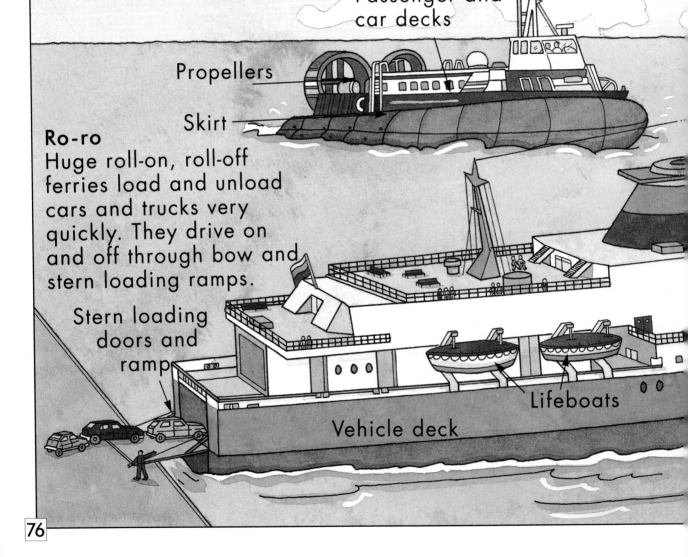

Passenger and car decks

Propellers

Skirt

Stern loading doors and ramp

Lifeboats

Vehicle deck

Hydrofoil
This ferry skims across the water on small, very strong wings called foils.

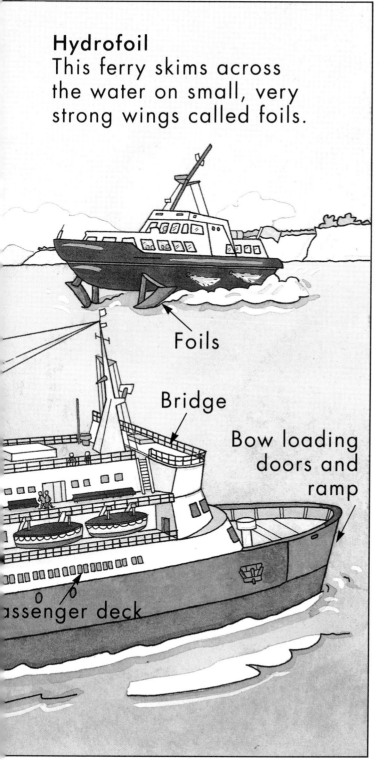

Skirt
The hovercraft's skirt fills with air. It can ride on air over ground as well as on waves.

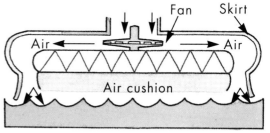

Foils
Foils are like aeroplane wings. Hydrofoils have to get up speed before they can 'take off'.

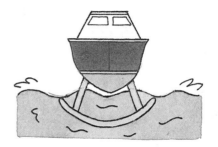

Fishing

Fishing at sea can be a dangerous job. Fishermen often risk their lives in storms and freezing weather.

Fun Fact

Machinery on big factory ships can prepare fish in minutes.

Fish factory ship
The catch is dropped onto a conveyor belt. Inside the ship the fish is cleaned, packed and frozen. It is often stored for weeks in a large refrigerated hold.

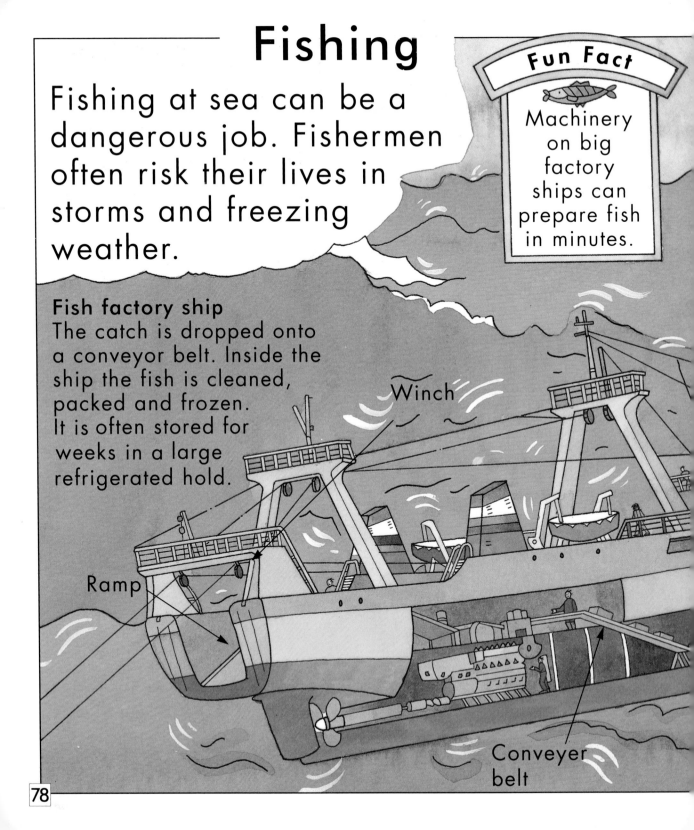

Winch

Ramp

Conveyer belt

Trawler
Trawlers are usually small. The fish is packed in ice and only keeps fresh for about 14 days.

Refrigerated store

Bringing in the catch

A net called a **trawl** is dragged through the water.

A **winch** hauls the trawl full of fish up the stern ramp.

The crew prepares the fish in safety on the **lower deck**.

Floating hotels

Cruise ships are big floating hotels. Passengers travel in comfort to interesting places all over the world. There is plenty to do on board.

Bridge The captain and officers control the ship from the bridge.

Stabilizers Small fins, called stabilizers, help to stop the ship rolling from side to side.

Propellers Powerful engines turn huge propellers which drive the ship through the water.

Fun Fact

Passenger liners take about 3 days to cross the Atlantic. Concorde can fly from London to New York in under 4 hours!

Cargo carriers

Some ships are specially built to carry enormous amounts of cargo. They can carry loads for long distances, from country to country.

Container ship
Containers arrive at the port on trucks and trains. High-speed lifting cranes load them on to the container ship.

Moving crane

Containers

Supertanker

This supertanker carries over 500,000 tonnes of oil in rows of tanks in its hull.

Tank hatches

Supertankers are too big for most harbours. Oil is loaded and unloaded in deep water through **pipes** into **smaller tankers** which carry the oil ashore.

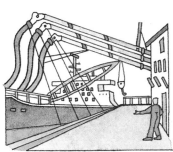

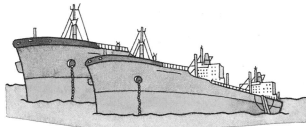

Containers are big boxes filled with cargo that stack neatly together.

Fun Fact

Supertankers are so long that sometimes the crew ride bicycles to get from one end of the deck to the other!

Warships

Different kinds of warships are designed to fight on the water, or underwater, or to help planes fight in the air.

Aircraft carrier
Fighter planes and helicopters take off and land on aircraft carriers.

Destroyer
Destroyers and cruisers armed with guns and missiles fight on the water.

Submarines

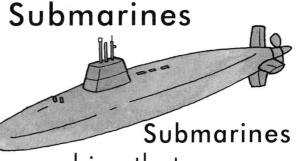

Minesweeper
Underwater explosives called mines are found and destroyed by minesweepers.

Submarines are ships that can travel under water. They carry special missiles, called torpedoes, which can be fired under water.

The **periscope** is the submarine's eye. It is a long tube with mirrors that reflect what is above the surface of the water.

Ships for special jobs

Dredgers are used to move mud off the seabed near to the shore so ships do not get stuck. They are also used to collect building materials such as rocks.

Fun Fact

Mud dumped out at sea is nearly always washed back in again!

Buckets

Chute

Barge

Icebreakers ride up onto the ice. The heavy bow breaks the ice. Then the extra-wide hull clears a wide path.

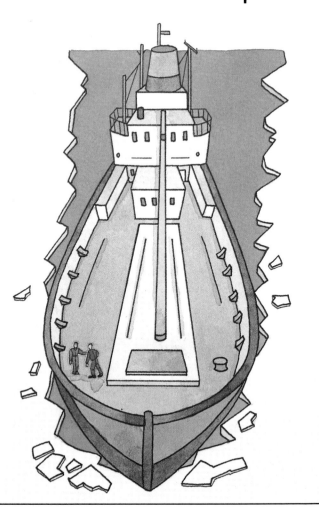

Special features

Dredger **buckets** move round on a ladder scooping up mud. It slides down chutes into barges and is dumped further out to sea.

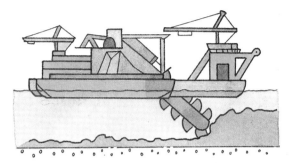

Some dredgers use pipes to suck up soft mud.

Sailing

Sailing ships, boats and yachts catch the wind in their sails to move them along.

- Sail
- Forestay
- Jib
- Bow sprit
- Shroud

Dinghy
One sail and a good wind is enough to speed this dinghy along.

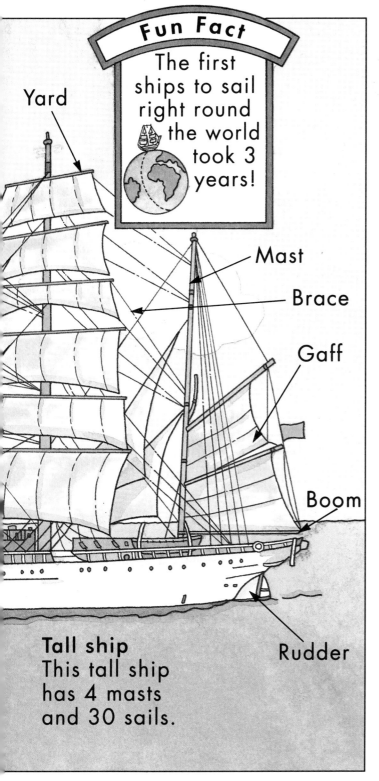

Fun Fact
The first ships to sail right round the world took 3 years!

Yard
Mast
Brace
Gaff
Boom
Rudder

Tall ship
This tall ship has 4 masts and 30 sails.

Sails

Sails are attached to a tall pole called a **mast**.

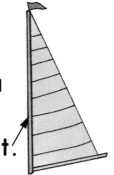

The sail is stretched out onto a **boom** which can swing round to catch the wind.

The crew control the sail with a rope called a **sheet**.

Speeding

Fast boats are built for racing and having fun, and also rescuing.

Racing boat
Racing boats have pointed bows and powerful engines.

Cruiser
You can eat, sleep and live in comfort on board a cruiser.

Lifeboat
Lifeboats are specially designed to stay afloat in very rough seas.

Motorboat
A motorboat can tow a waterskier.

Outboard engines are fitted on the outside of a boat. They can be taken on and off.

Inboard engines are built into a boat.

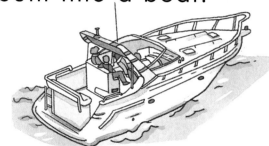

Jet skis are like motorbikes on water.

Oars and paddles

Boats without engines or sails need human power to move them along.

Rowing boat

Rowing eights
Rowing eights are used for racing. The cox steers and shouts instructions to the crew.

Kayak
A kayak is a very light canoe.

Methods

One paddle
Canoeists hold one paddle with both hands. They push it into the water on either side of the canoe and pull back.

Two oars can be used
on either side of a rowing boat.

One oar
Using one oar at the back of the boat is called sculling.

Finding the way

There are no roads and signposts at sea to help sailors. These are some of the things they use to help them find their way.

Navigation lights
A red port light and a green starboard light show the direction a ship is sailing in.

Fun Fact
Sailors have always used the position of the stars and planets to find their way.

Satellite
Satellites send out signals that help ships to work out exactly where they are.

Lighthouse
Lighthouses mark harbour entrances and warn ships of dangerous rocks and tides.

Buoy
Floating marker buoys mark rocks and other dangers.

Radar shows where other ships and land are so ships can sail at night or in fog without colliding.

Charts are sea maps showing where dangerous rocks and sandbanks are.

A **compass** points to the North. Ships use them to work out which way to go.

Did You Know?

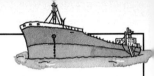

When at sea people on different ships can understand each other from a distance using the International Code of Signals. The ships raise flags that people on other ships can recognise whatever language they speak.

Now yachts can sail right round the world in less than 8 months. The first journey round the world took 3 years.

Ships have been built in strange shapes to see which shape sails better. One ship was made of three small ships hinged together in a zigzag shape.

Written by Sally Hewitt and Nicola Wright
Designed by Chris Leishman
Illustrated by Rachael O'Neill
Edited by Dee Turner, Nicola Wright, Fiona Mitchell
Series concept: Tony Potter
Consultants: Captain Ian Evans, Andrew Woodward, B Eng (Hons), Robin Wright, M.M., M.A.
Cover design: Deborah Chadwick
Colour separations by Advance Laser Graphic Arts (International) Ltd Hong Kong, RCS Graphics Ltd Leeds, Scan Trans Singapore.
Printed by G Canale & Co SpA, Italy

This book was created by Zigzag Publishing Ltd, The Barn, Randolph's Farm, Brighton Road, Hurstpierpoint, West Sussex, BN6 9EL

First published in 1995 by Zigzag Publishing Ltd
Copyright © 1995 Zigzag Publishing Ltd

All rights reserved. No part of this publication may be reproduced, stored in a retrieval system or transmitted by any means, electronic, mechanical, photocopying or otherwise, without the prior permission of the publisher.

ISBN 1 85993 004 2

10 9 8 7 6 5 4 3 2 1

By
Josie Akin

MY JUBILEE YOUR JUBILEE

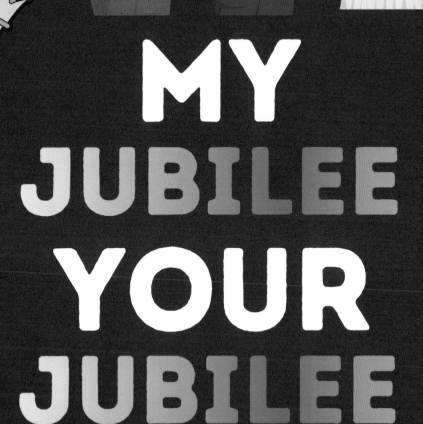

My Name Is _____

© 2022 Josie Akin

This is a work of fiction and is created for entertainment purposes only.

All rights reserved. No part of this publication may be reproduced, distributed or transmitted in any form or means without prior permission of the publisher.

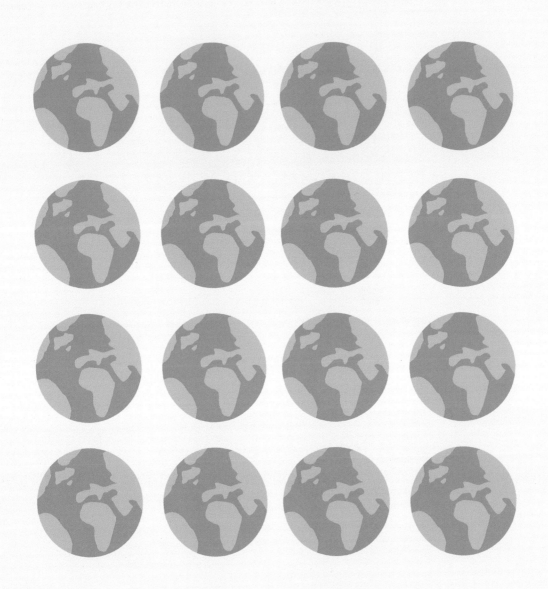

It was a lovely June morning
The day of the Platinum Jubilee
People were coming from all over the world
To see Her Majesty The Queen

"I cannot wait for my guests to arrive"
The Queen joyfully declared
"May I have a look the guest list?"
"One must be prepared".

The Queen glanced at her list
And wow was she in for a treat
A guest from every country of her Commonwealth
Lots of interesting people to meet

It was 3pm
When the first guest arrived
It was the Pretty Princess of India
Who had a twinkle in her eyes.

"Hello, Your Majesty"
The Princess of India beamed
"Congratulations on 70 years!"
"You have been a wonderful Queen!"

Before the Queen could reply
There was a knock on the door
It was the Princess of Zambia
"Good day, Your Highness, we have met before".

The Queen was pleased
To see the Princesses of Zambia and India
A few moments later
Arrived the Prime Minister of Australia.

Next to arrive
Were the Princes of Malta and Pakistan
With lots of delicious food
They had brought from their land

Before the Queen knew it
Her guests had filled the room
From the lovely ladies of Barbados and Samoa
And the kind King of Cameroon.

"Congrats again!" Said the Prince of Pakistan
"70 glorious years on the British Throne!"
"Please share an interesting story about your reign
It would be very lovely to know"

"Okay everyone" The Queen replied
"Please gather around
I will tell you all about the time
we nearly lost the crown.

It was a rather long time ago
The night before my Silver Jubilee
I was young and full of life
And had served 25 years as Queen.

Parades were due to take place
Decorations filled Palace grounds
The only thing that was missing
Was my favourite gold and purple crown.

We searched the gardens
And the royal rooms
We even went to Windsor castle
And searched there too

The Jubilee celebrations were about to start
And the crown was still nowhere to be seen
"We have searched everywhere" I cried
Husband replied "What about the Corgi?".

My husband and I walked over to the dogs room
And opened the kennel door
And there was my beautifully regal corgi
Wearing a crown filled with fur.

So I found my crown
And enjoyed the Jubilee celebrations
We took lots of amazing pictures
For the glorious occasion.

"That ends my story
Thanks for helping me celebrate my Platinum Jubilee!"
"I am just as proud of the UK and Commonwealth
As when I first became Queen".

The representatives of the Commonwealth
Enjoyed the Queen's story
70 years as the Queen of Britain
The Land of Hope and Glory.

Printed in Great Britain
by Amazon

Mae Pawb yn Myna ar Saffari
Taith gyfrif drwy Tanzania

*I'm hwyrion – Tim, Gibson, Billy, Joe, Bennett –
Ac i blant Ysgol Gynradd Farmingville* – L. K.

I Helen a Lucy, gyda chariad – J. C.

Hoffai Barefoot Books diolch i Dr. Michael Sheridan,
Athro Cynorthwyol Preswyl mewn Cymdeithaseg ac Anthropoleg yng Ngholeg Middlebury, Vermont,
am ei gymorth caredig gyda chyfieithu ac ynganu'r Swahili

Hawlfraint y testun © 2003 gan Laurie Krebs
Hawlfraint y lluniau © 2003 gan Julia Cairns
Addaswyd gan Mererid Hopwood (cerddi) ac Elin Meek (gweddill y testun) 2007
Mae Laurie Krebs a Julia Cairns wedi datgan eu hawl foesol i gael eu cydnabod fel awdur
ac arlunydd y llfr hwn.

Cyhoeddwyd gyntaf yn y Deyrnas Unedig yn 2003 gan Barefoot Books Ltd, 124 Walcot Street, Bath, BA1 5BG.
Teitl gwreiddiol: We All Went on Safari
Argraffiad Cymraeg cyntaf yn 2007 gan Llyfrau Barefoot (Cymru) Cyf.,
Suite 112, 61 Wellfield Road, Cardiff, CF24 3DG.
Cedwir pob hawl. Ni chaniateir atgynhyrchu unrhyw ran o'r chyoeddiad hwn, na'i gadw mewn cyfundrefn adferadwy,
na'i drosglwyddo mewn unrhyw ddull na thrwy unrhyw gyfrwng, electronig, electrostatig, tâp magnetig, mecanyddol,
llungopïo, recordio nac fel arall, heb gantiatâd ymlaen llaw gan y cyhoeddwyr.

Cysodwyd y llyfr hwn yn Legacy
Paratowyd y darluniau mewn dyfrlliw

Dylunio graffeg gan Louise Millar, Llundain
Atgynhyrchu lliw gan Grafiscan, Verona
Argraffwyd a rhwymwyd yn China gan Printplus Ltd.

Argraffwyd y llyfr hwn ar bapur cwbl ddi-asid.

ISBN 0-9552659-1-6   978-0-9552659-1-4

Mae cofnod ar gyfer y llyfr hwn ar gael o'r Llyfrgell Brydenig

1 3 5 7 9 8 6 4 2

Mae Pawb yn Mynd ar Saffari

Taith gyfrif drwy Tanzania

Ysgrifennwyd gan Laurie Krebs
Addaswyd gan Mererid Hopwood ac Elin Meek
Darluniau gan Julia Cairns

Llyfrau Barefoot
Dathlu Celf a Storïau

Mae pawb yn mynd ar saffari
â hithau'n doriad gwawr.

'Dacw un llewpart lliwgar!' medd Arusha, 'ac O! mae ei smotiau'n fawr.'

moja 1

Mae pawb yn mynd ar saffari
a'r gwlith yn oer dan droed.

'Ust! Dau estrys!' medd Mosi,
'y rhyfeddaf a fu erioed!'

mbili **2**

Mae pawb yn mynd ar saffari heibio'r acasia hen.

'Tri jiraff yn joyo!' meddai Tumpe gyda gwên.

tatu **3**

Mae pawb yn mynd ar saffari
at ddŵr hynafol y llyn,

lle mae'r llewod yn syllu'n llonydd.
'Pedwar!' medd Mwambe yn syn.

nne 4

Mae pawb yn mynd ar saffari
a gweld adar yn dawnsio'n y dŵr.

'Dacw'r hipos hapus,' medd Akeyla, 'edrychwch! Rwy'n gweld pump, rwy'n siŵr.'

tano 5

Mae pawb yn mynd ar saffari
i gael golwg ar deulu'r gnŵ.

'Rwy'n gweld chwech, yn wir,' medd Watende, 'mae 'na chwech! Oes! Dacw nhw!'

sita
6

Mae pawb yn mynd ar saffari ac mae bellach yn ganol dydd.

'Saith sebra urddasol,' medd Zalira,
'yn mwynhau yn yr awel rydd.'

saba 7

Mae pawb yn mynd ar saffari
i'r Serengeti i bwyso ar glwyd.

'Wyth baedd dafadennog!' medd Suhuba, 'yn frwnt ac yn frown ac yn llwyd.'

nane 8

Mae pawb yn mynd ar saffari
at y goeden am gysgod rhag yr haul.

'Naw mwnci!' medd Doto'n llawn cyffro, 'naw mwnci yn dawnsio'n y dail.'

tisa 9

Mae pawb yn mynd ar saffari
i'r llanerch garegog a chlyd.

'Deg eliffant llwyd,' meddai Bodru, 'yn bwyta'r brigau i gyd.'

kumi 10

Mae pawb yn mynd ar saffari
â hithau yn hwyrnos o ha',

i eistedd yng ngwres y fflamau
a dweud: 'Tan tro nesa - nos da!'

Anifeiliaid Tanzani

Llewpard – chui *(tsŵ-i)*
Mae'r llewpard yn aml yn cario ei ysglyfaeth i gangen uchel, lle gall fwyta a chysgu'n ddiogel. Mae'n anodd iawn ei weld yn ei guddfan – ond weithiau mae ei gynffon smotiog yn hongian yn y golwg.

Llew – simba *(sim-ba)*
Y llew benyw sy'n hela dros y cnud o lewod. Weithiau mae hyd at un deg tri o aelodau yn y teulu.

Estrys – mbuni *(m-bw-ni)*
Mae estrys yn dalach na'r rhan fwyaf o chwaraewyr pêl fasged proffesiynol, tua 7 i 8 troedfedd. Maen nhw'n rhedeg yn gyflym iawn hefyd!

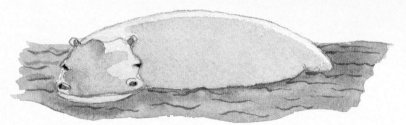

Hipopotamws – kiboko *(ki-bô-co)*
Mae hipos yn gorwedd yn y dŵr yn ystod y dydd. Maen nhw'n plygu eu clustiau a chau eu ffroenau. Felly dydy eu croen nhw ddim yn sychu yn yr haul.

Jiráff – twiga *(twî-ga)*
Mae tafodau hir tua 45cm gan jiraffod a gwefusau uchaf fel sbwng. Felly gallant fwyta o gwmpas pigau eu hoff fwyd, y goeden acasia.

Gnw Cyffredin – nyumbu (*naiym*-bw)

Mae gnw'n edrych fel cyfuniad o lawer o anifeiliaid gwahanol. Mae ganddo ben ychen, mwng ceffyl, cyrn byfflo a barf gafr.

Sebra – punda milia (***pwn**-da mi-**lî**-a*)

Mae patrwm du a gwyn arbennig gan bob sebra, yn union fel mae gan bob person olion bysedd. Mae'r patrwm yn guddliw i'r sebra ar doriad gwawr a phan fydd yr haul yn machlud. Dyma pryd mae llewod yn hela.

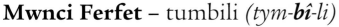

Mochyn dafadennog – ngiri (***ngî**-ri*)

Mae teuluoedd o foch dafadennog yn trotian yn rhes y tu ôl i'w gilydd – y fam yn y blaen a'r moch bach y tu ôl iddi. Mae cynffon pob un yn codi'n syth i'r awyr.

Mwnci Ferfet – tumbili (*tym-**bî**-li*)

Mae mwncïod ferfet bach yn cael reid gyda'u mam drwy ddal ei ffwr a lapio'u cynffonau am ei chefn.

Eliffant – tembo (***tem**-bo*)

Mae'r fam yn gofalu am eliffant bach yn dyner iawn. Mae'n ei gadw oddi tani rhwng ei choesau. Hefyd, mae'n gwneud yn siŵr ei fod yn ddiogel yng nghanol y grŵp pan fydd y gyrr ar daith.

Llwyth y Maasai

Mae llwyth Maasai Dwyrain Affrica'n byw yn yr ardal yng ngogledd Tanzania sy'n ffinio â de Kenya. Mae llawer o deuluoedd yn byw gyda'i gilydd mewn pentrefi bach. Maen nhw'n adeiladu eu cartrefi o fwd, prennau, porfa a thail gwartheg. Maen nhw'n bugeilio gyrr mawr o wartheg gyda'i gilydd. Dyma eu prif waith. Pan fydd y borfa'n dda a digon o laswellt, bydd y bobl yn aros lle maen nhw. Pan fydd y tir yn sychu a'r tymhorau'n newid, bydd y grŵp yn symud ymlaen i ddod o hyd i ddŵr ffres a phorfeydd newydd i'r gwartheg.

Mae'r Maasai'n bobl falch. Maen nhw'n dal ac yn olygus yn eu clogynnau coch cyfoethog. Mae dynion a menywod yn eu haddurno'u hunain â chlustdlysau a neclisau gleiniau hardd. Weithiau mae'r dynion yn gwisgo'u gwallt yn smart ac yn gwisgo penwisgoedd coeth. Mae'r menywod fel arfer yn eillio eu pennau ac yn gwisgo coleri gwyn siâp cylch sy'n siglo'n rhythmig wrth iddyn nhw symud.

Mae'r Maasai wedi byw yng nghanol bywyd gwyllt Dwyrain Affrica ers miloedd o flynyddoedd. Ond wrth i'r byd newid mor gyflym, maen nhw'n gorfod brwydro i gadw eu ffordd o fyw. Nhw yw un o'r diwylliannau bugeilio olaf ar y ddaear.

Enwau Swahili

Pan fydd rhieni o lwyth y Maasai'n dewis enw i'w plentyn, byddan nhw'n aml yn dewis enw sydd ag ystyr arbennig. Y gobaith yw y bydd y baban yn tyfu i fod fel yr enw.

ARUSHA (b) (*a-**rw**-sia*) – annibynnol, creadigol, uchelgeisiol

MOSI (g) (***mo**-si*) – amyneddgar, cyfrifol, yn caru ei deulu a'i gartref

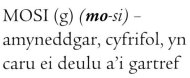

TUMPE (b) (***twm**-pei*) – cyfeillgar, doniol, arweinydd a threfnydd

MWAMBE (g) (***mwam**-bei*) – taclus, tawel, dyn busnes da

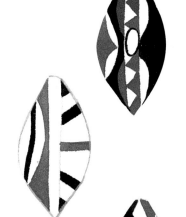

AKEYLA (b) (*a-**cei**-la*) – yn hoff o fyd natur a'r awyr agored

WATENDE (g) (*wa-**ten**-dei*) – sensitif, hael, creadigol

ZALIRA (b) (*sa-**lî**-ra*) – llawn cydymdeimlad, tawel, cyfeillgar

SUHUBA (g) (*sw-**hw**-ba*) – deallus, talentog, serchog

DOTO (g/b) (***do**-to*) – hael, serchog, parod i helpu

BODRU (g) (***bo**-drw*) – gweithgar, yn cymryd amser i orffen popeth mae'n ei ddechrau

Ffeithiau am Tanzania

Tanzania yw'r wlad fwyaf yn Nwyrain Affrica. Mae bron cymaint â Ffrainc.

Mynydd Kilimanjaro yw'r mynydd uchaf yn Affrica. Mae'n 5,895 metr (19,340 troedfedd) o uchder.

Llyn Victoria, yn y gogledd, yw'r llyn mwyaf ond un yn y byd.

Cyn 1961, enw'r wlad oedd Tanganyika. Dyma'r enw o hyd ar un o'r llynnoedd, Llyn Tanganyika. Erbyn hyn mae Tanzania'n cynnwys Tanganyika ac ynys Zanzibar.

Mae mwy na 100 llwyth yn byw yn Tanzania.

Ystyr Serengeti yw 'y gwastadedd diddiwedd'.

Hen losgfynydd wedi cwympo yw Crater Ngorongoro. Unwaith roedd yn uwch na Mynydd Kilimanjaro. Erbyn hyn mae'n edrych fel powlen ddwfn.

Weithiau mae Ceunant Olduvai'n cael ei alw'n grud y ddynolryw. Daeth esgyrn dynion cynnar i'r golwg yno.

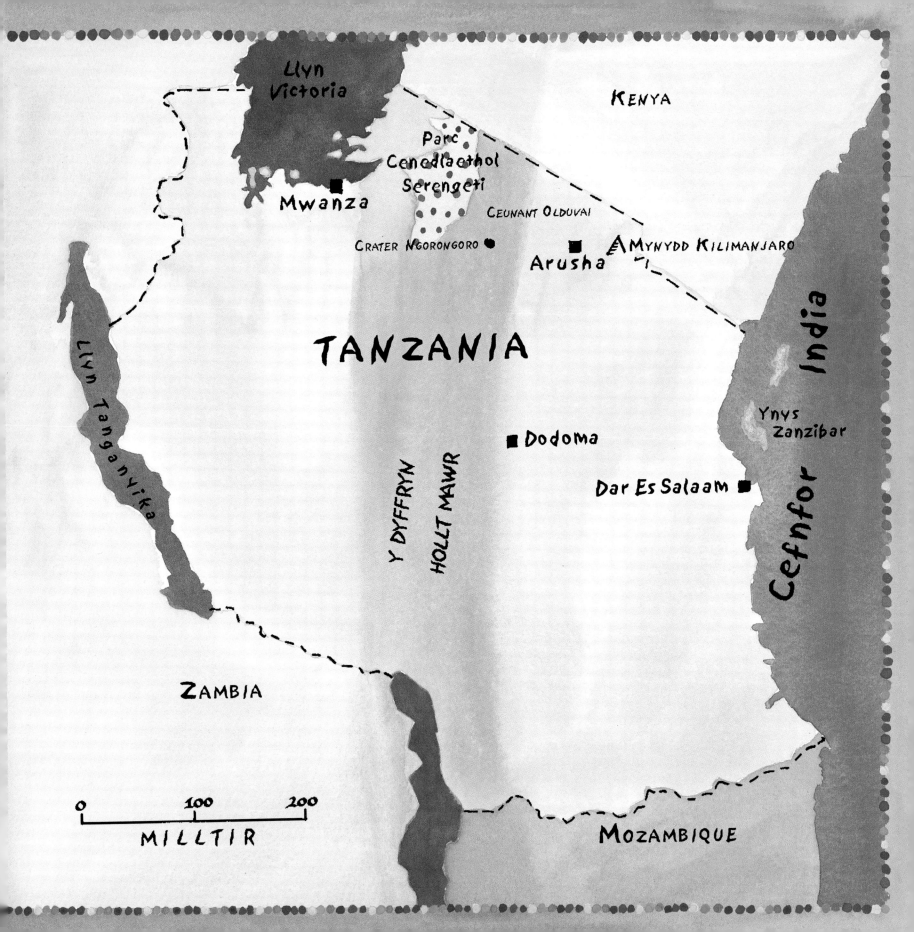

Cyfrif mewn Swahili

1 •	**6** • • • • • •
moja	sita
(mo-ja)	*(sî-ta)*
un	chwech
2 • •	**7** • • • • • • •
mbili	saba
(m-bî-li)	*(sa-ba)*
dau	saith
3 • • •	**8** • • • • • • • •
tatu	nane
(ta-tw)	*(na-nei)*
tri	wyth
4 • • • •	**9** • • • • • • • • •
nne	tisa
(n-nei)	*(tî-sa)*
pedwar	naw
5 • • • • •	**10** • • • • • • • • • •
tano	kumi
(ta-no)	*(cŵ-mi)*
pump	deg